This book belongs to

Let's Learn the Number

zero

zero

O   O   O   O   O

Let's Learn the Number

ONE

one

1   1   1   1

Let's Learn the Number
TWO

two

2 2 2 2 2 2

Let's Learn the Number

THREE

three

3 3 3 3 3 3

Let's Learn the Number
FOUR

four

Let's Learn the Number

FIVE

five

5   5   5   5   5

Let's Learn the Number

SIX

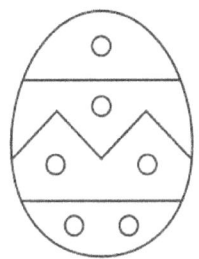 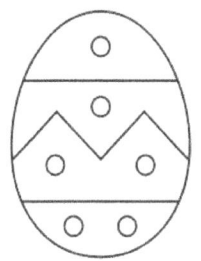 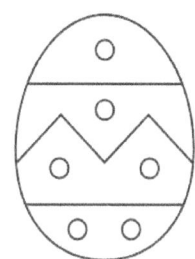

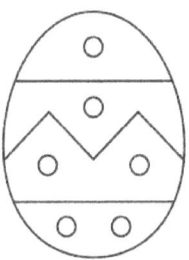 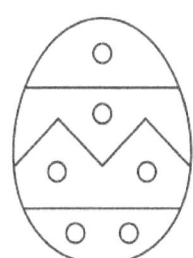 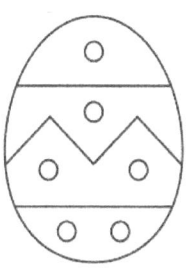

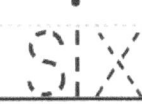

six

6   6   6   6   6   6

Let's Learn the Number
SEVEN

seven

7

Let's Learn the Number

EIGHT

eight

8

Let's Learn the Number
NINE

nine

9

Let's Learn the Number

TEN

ten

10   10   10   10   10

Let's Learn the Number

ELEVEN

eleven

Let's Learn the Number
TWELVE

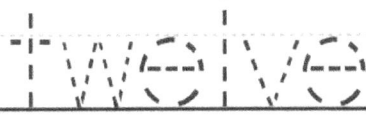

twelve

12   12

Let's Learn the Number
THIRTEEN

thirteen

13

Let's Learn the Number

FOURTEEN

fourteen

14

Let's Learn the Number

FIFTEEN

fifteen

15 15 15 15 15 15

Let's Learn the Number
SIXTEEN

sixteen

16   16   16   16   16

Let's Learn the Number
SEVENTEEN

seventeen

17 17 17 17 17

Let's Learn the Number

EIGHTEEN

eighteen

18 18 18 18 18 18

Let's Learn the Number

NINETEEN

nineteen

19 19 19 19 19 19

Let's Learn the Number
TWENTY

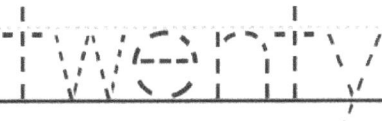

twenty

20 20 20 20 20

Let's Learn the Number

THIRTY

thirty

3 0 30 30 30 30

Let's Learn the Number
FORTY

forty

40   40   40   40

Let's Learn the Number

FIFTY

fifty

50

Let's Learn the Number
SIXTY

sixty

60   60   60   60   60

Let's Learn the Number
SEVENTY

seventy

70 70 70 70 70

Let's Learn the Number
EIGHTY

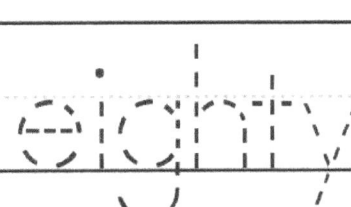

eighty

80   80   80   80   80

Let's Learn the Number
NINETY

ninety

90 90 90 90 90

Let's Learn the Number

ONE HUNDRED

one hundred

100 100 100 100 100

Color : *1 apple+ 1 banana+ 2 mangos*

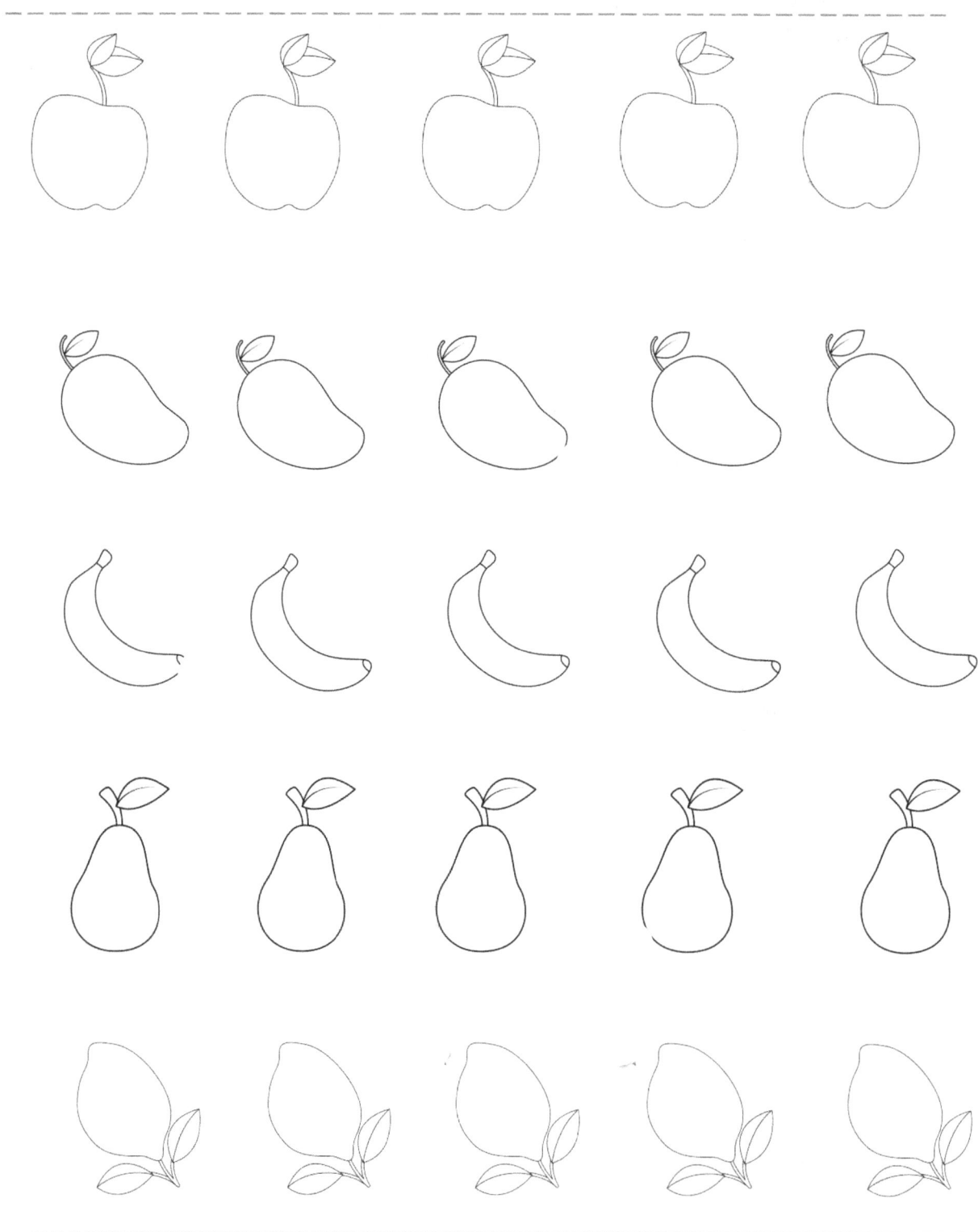

Color : *2 apples+1 lemon+ 3 bananas+4 avocados*

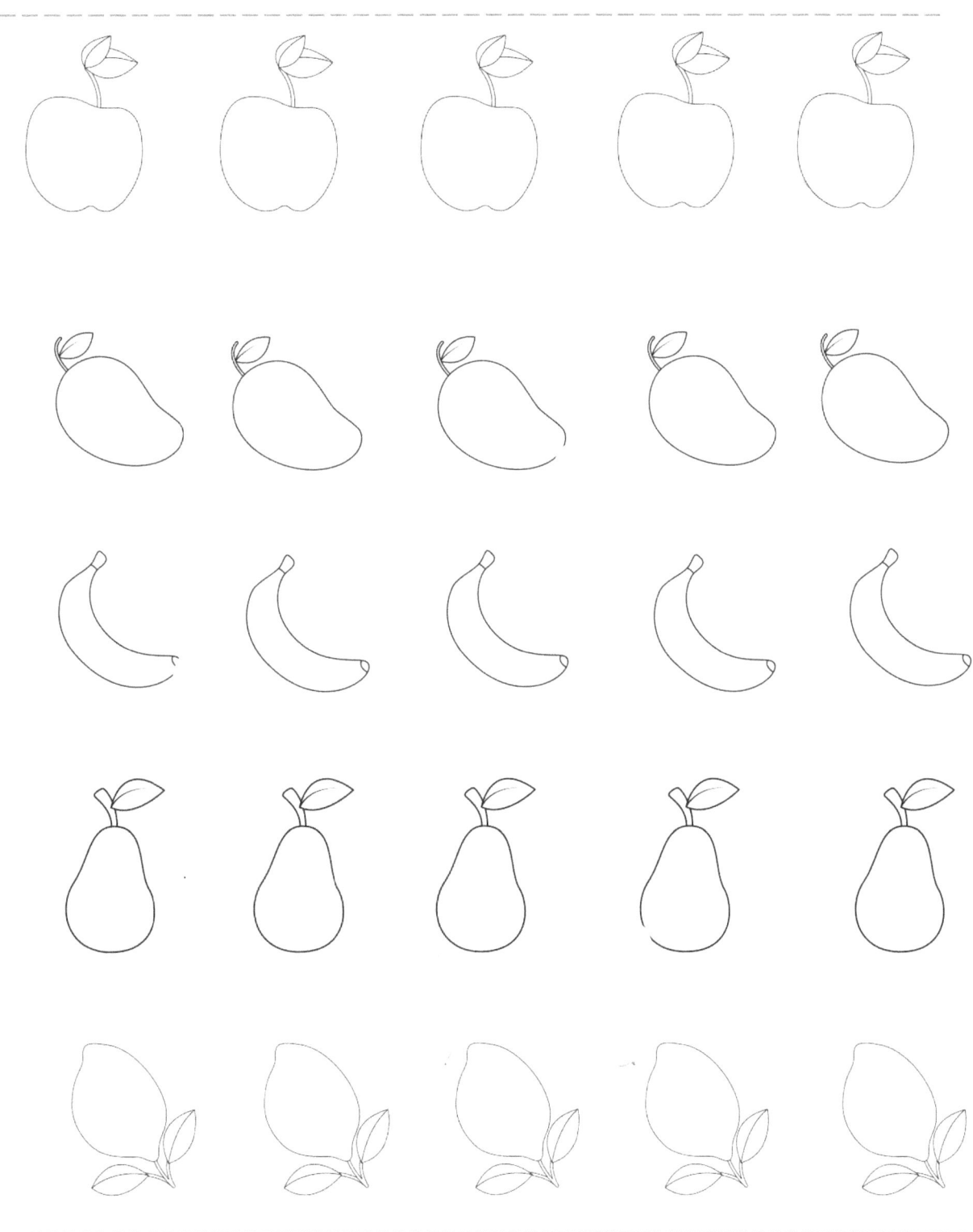

Color : *4 apples+ 2 avocados+ 3 bananas+ 1 lemon+ 3 mangos*

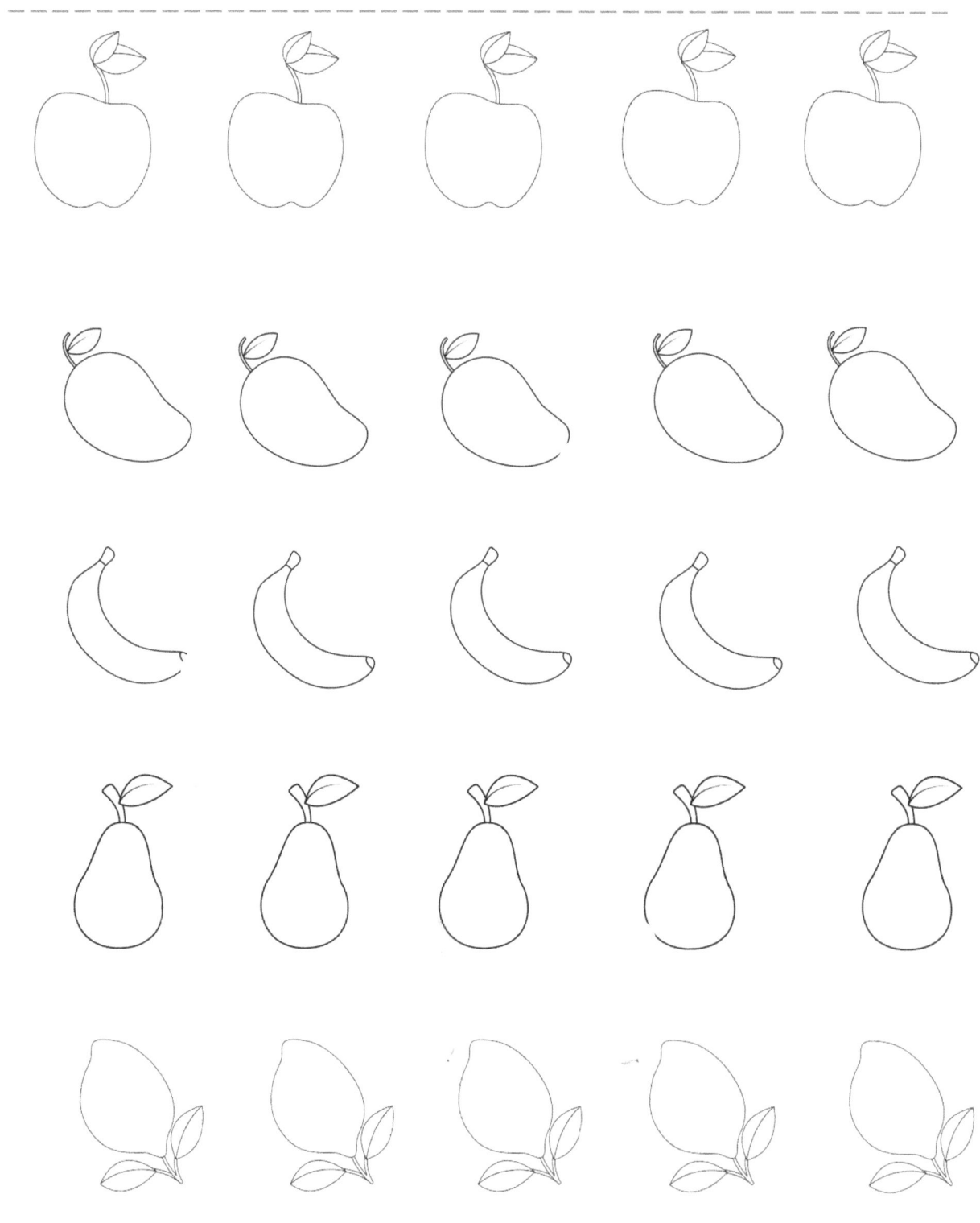

Color : 4 bananas+ 2 lemons+1 mango +2 apples+4 avocados

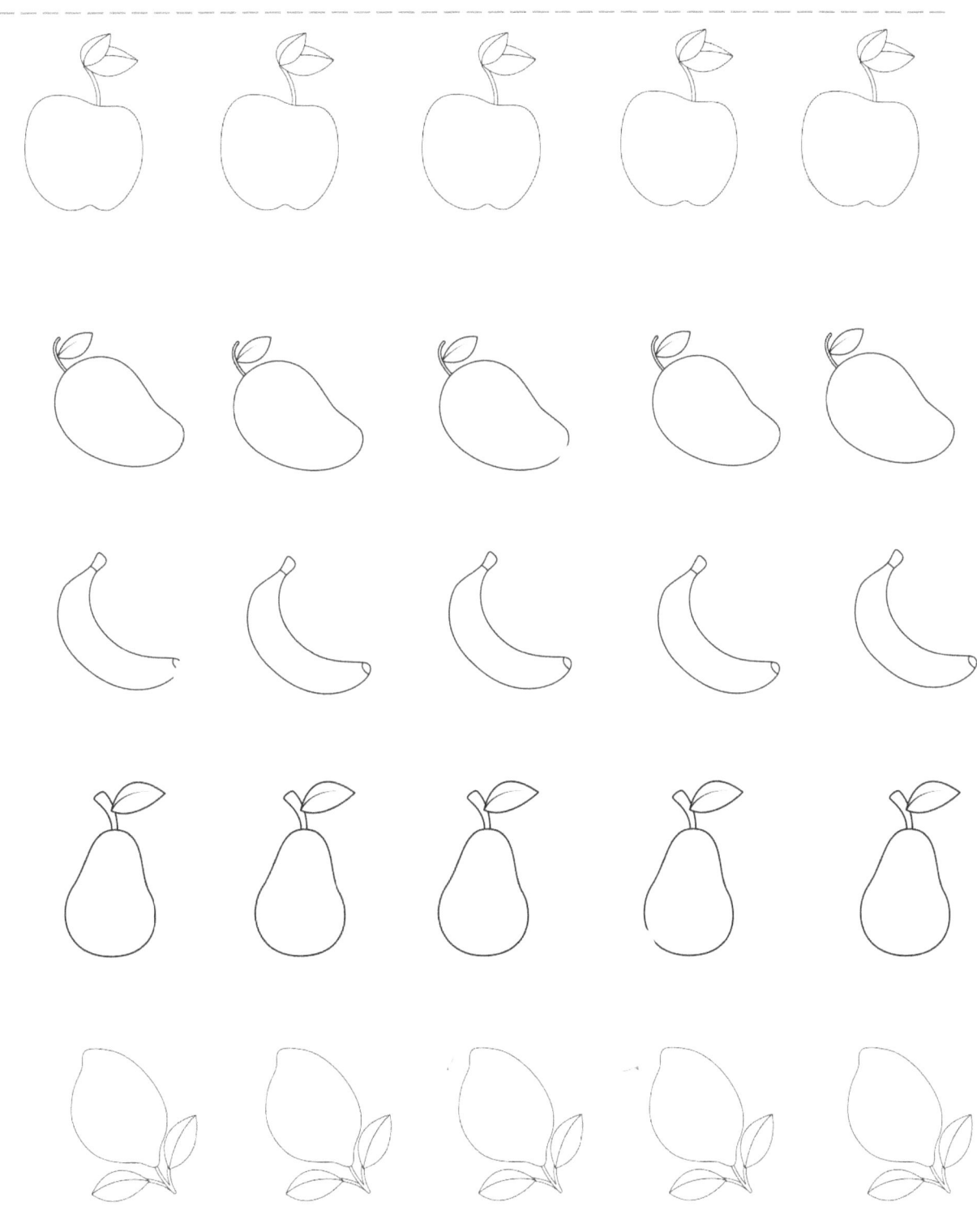

Color : *5 apples + 2 avocados +1 banana + 3 mangos + 2 lemons*

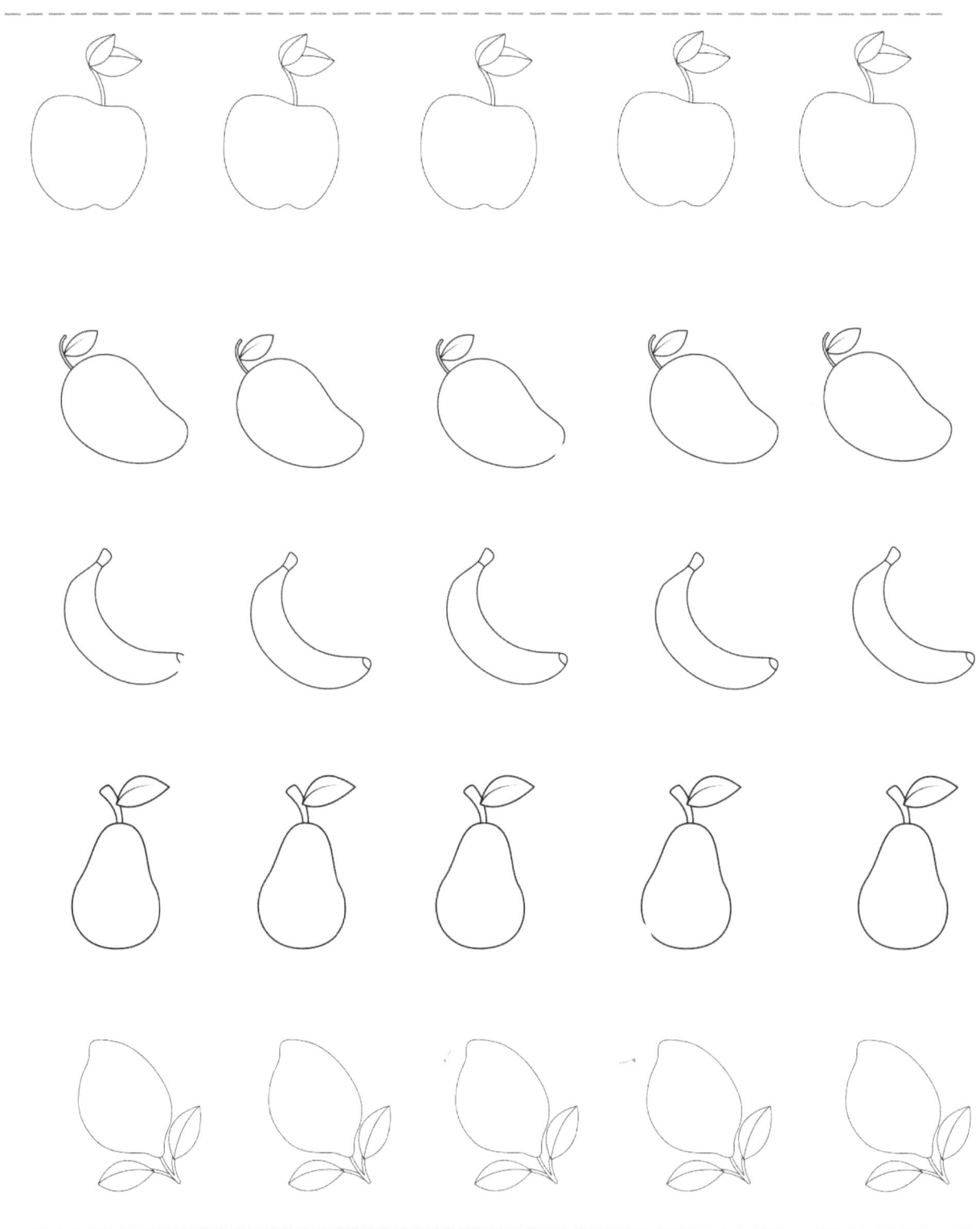

Color : *4 apples + 1 avocado 2 lemons + 4 bananas + 3 mangos*

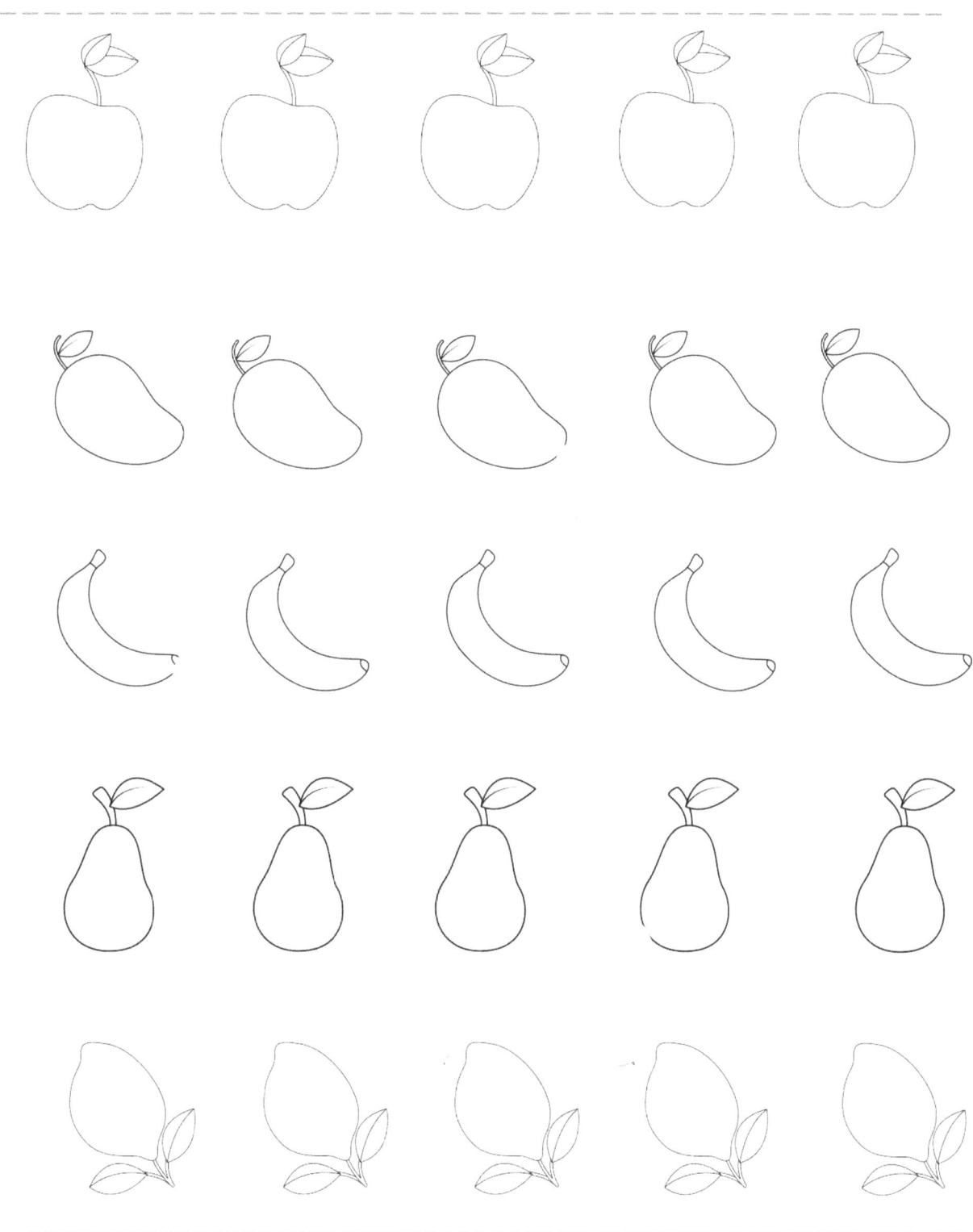

Color : *5 lemons + 2 mangos + 3 bananas + 4 avocados + 2 apples*

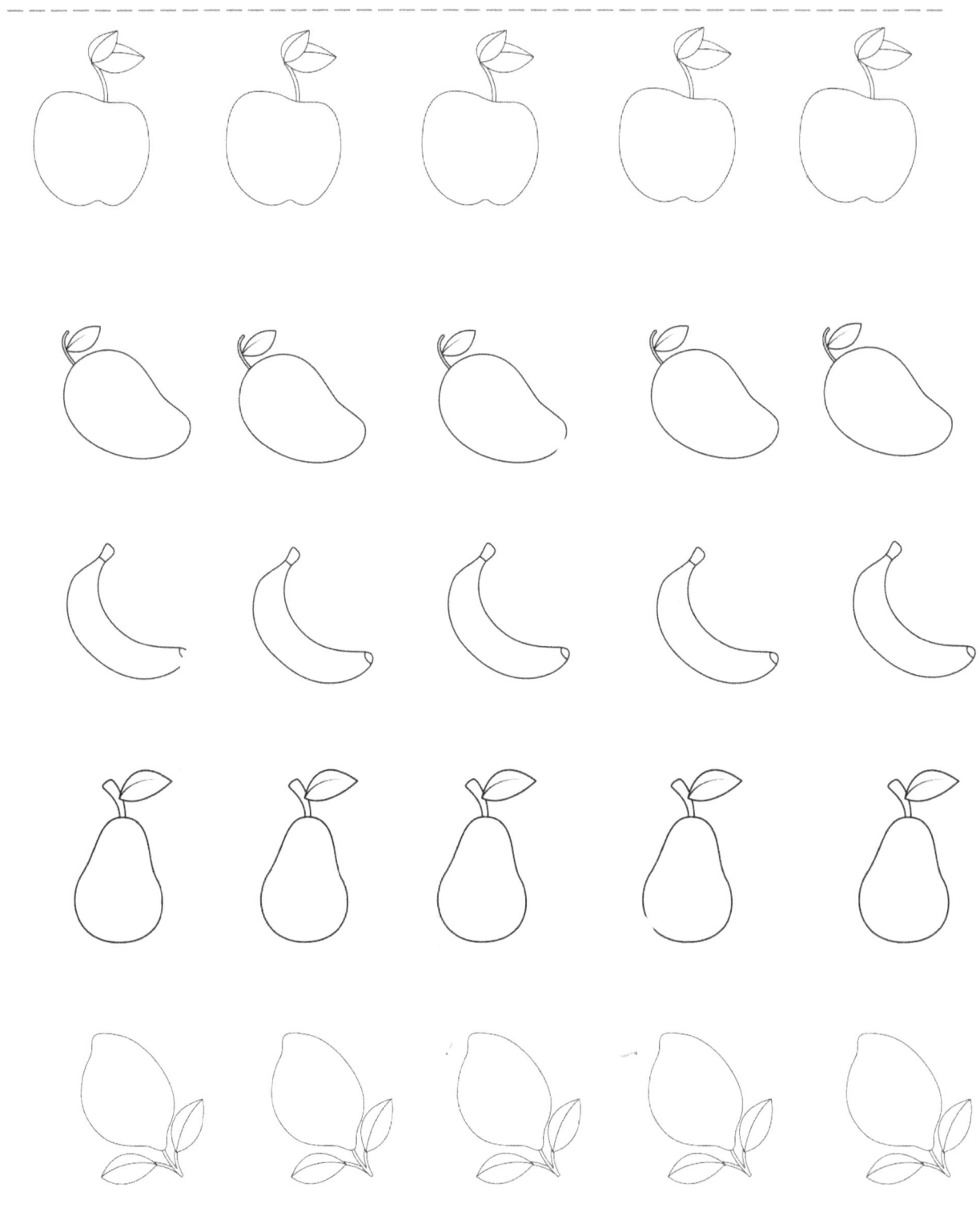

Color : 4 bananas + 3 mangos + 5 apples + 3 lemons 1 avocado

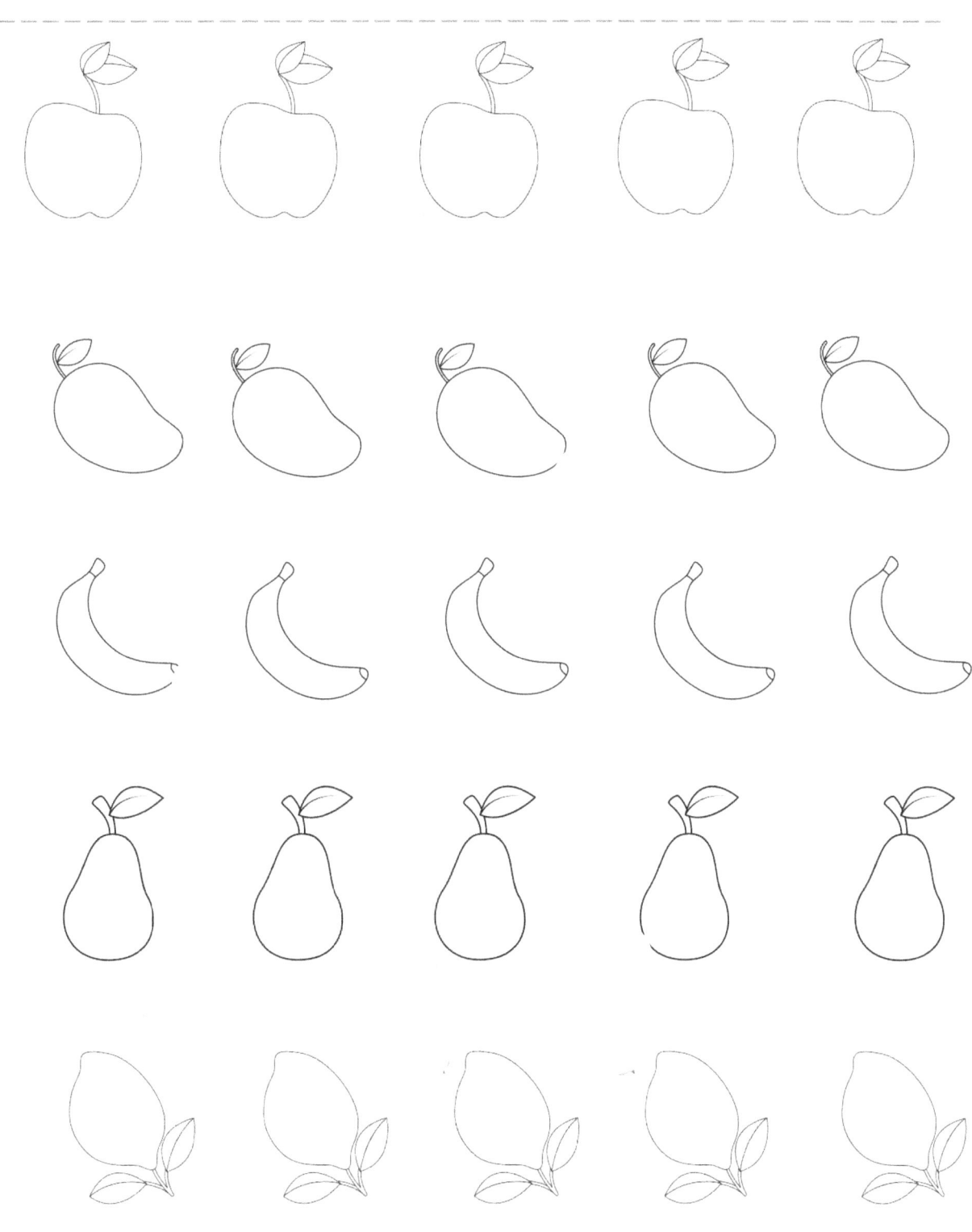

ADDITION

Add up the number of fruits and write your answer in the empty box below. Fill in the blank spaces to complete the name of the fruit.

FRUITS

WA _ ERM_LO_

_____

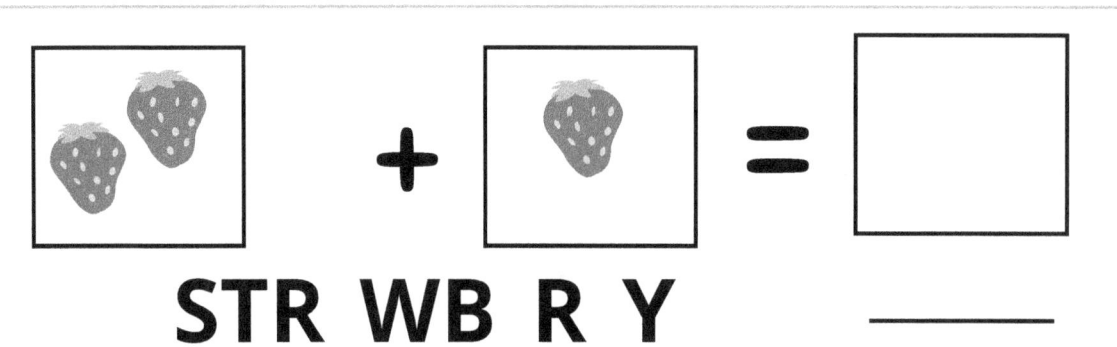

STR_WB_R_Y

_____

OR_NGE

_____

AP_LE

_____

SUBTRACTION

Name:

Grade & Section:

Age

Date:

Subtract the number of animals and write your answer in the empty box below. Fill in the blank spaces to complete the name of the animal.

ANIMALS

 -  =

BUTT_RF_Y _____

 -  =

MO_K_Y _____

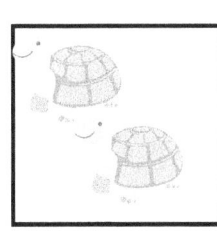 -  =

TO_TO_SE _____

 -  =

F_S_ _____

ADDITION

Name:

Grade & Section:

Age

Date:

Add up the number of balls and write your answer in the empty box below. Fill in the blank spaces to complete the name of the ball.

BALL

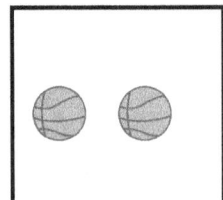 +  =

BASKET_A_L _____

 +  =

FOOT_A_L _____

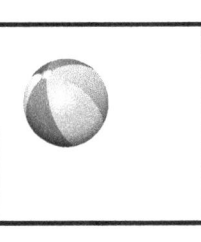 +  =

B_LL _____

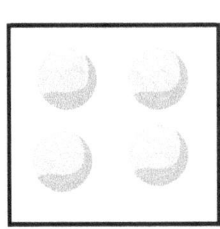 + 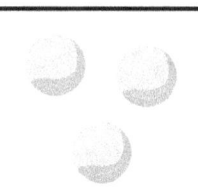 =

TE_N_S BA_L _____

ADDITION

Name:

Grade & Section:

Age

Date:

Add up the number of flowers and write your answer in the empty box below. Fill in the blank spaces to complete the name of the flower

FLOWERS

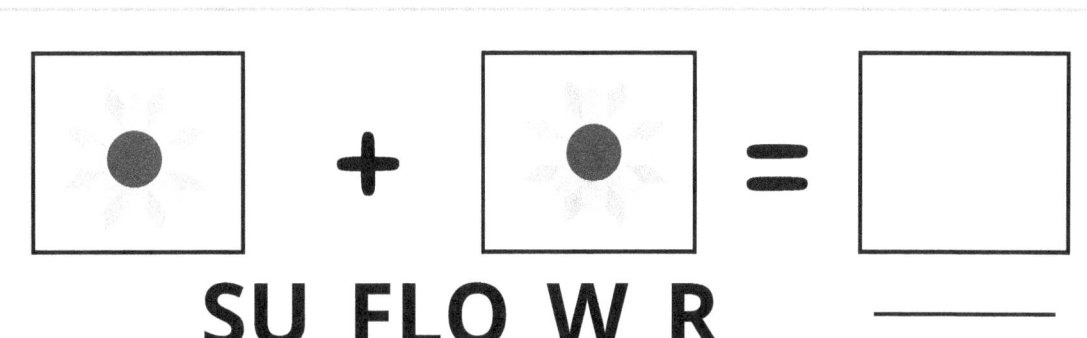

SU_FLO_W_R ____

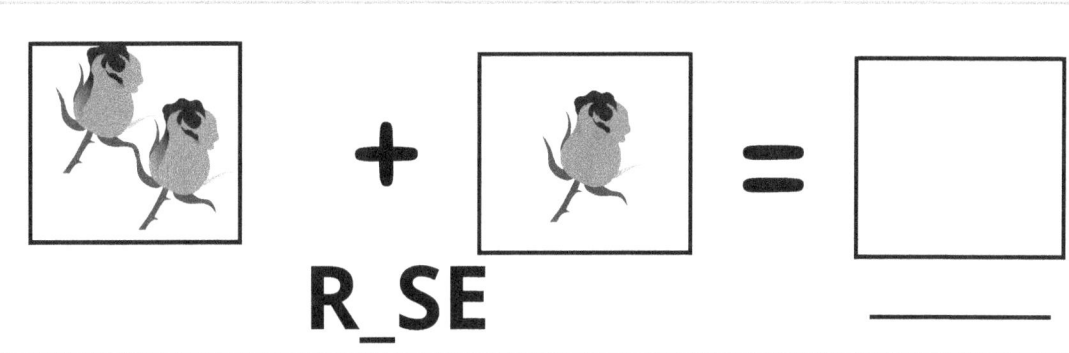

R_SE ____

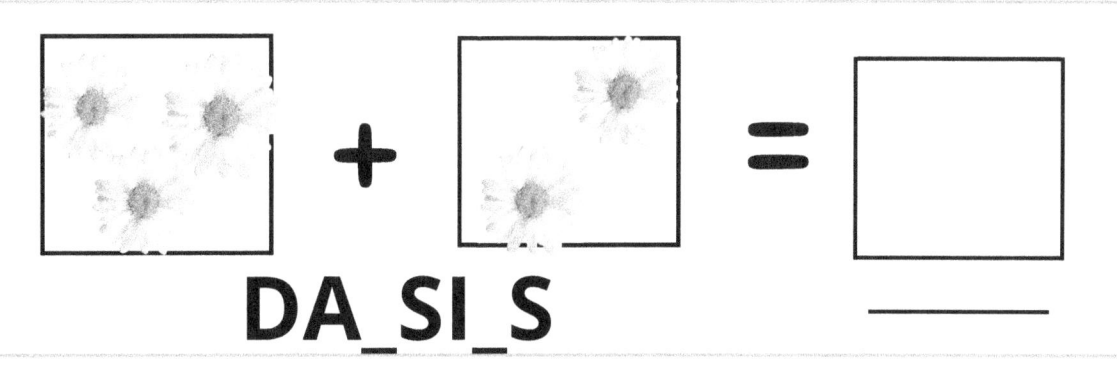

DA_SI_S ____

HIB_S_US ____

SUBTRACTION

Subtract the number of insects and write your answer in the empty box below. Fill in the blank spaces to complete the name of the insect

Name:

Grade & Section:

Age

Date:

INSECTS

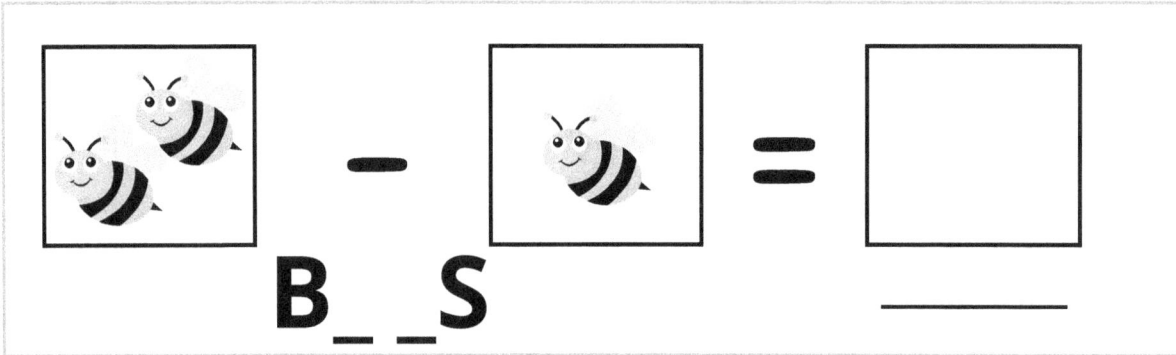

B _ _ S

_____

A _ T

_____

B _ ETL _

_____

GR _ SSHO _ P _ R _____

ADDITION

Name:

Grade & Section:

Age

Date:

Add up the number of furniture and write your answer in the empty box below. Fill in the blank spaces to complete the name of the furniture

FURNITURE

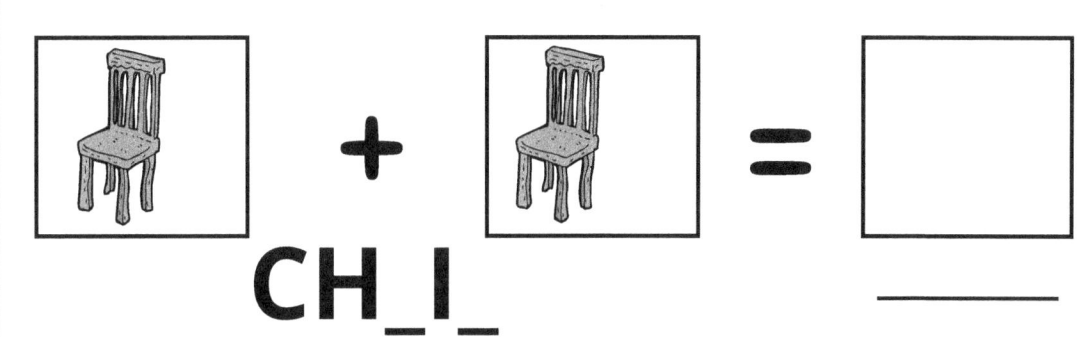

CH_I_

TA_LE

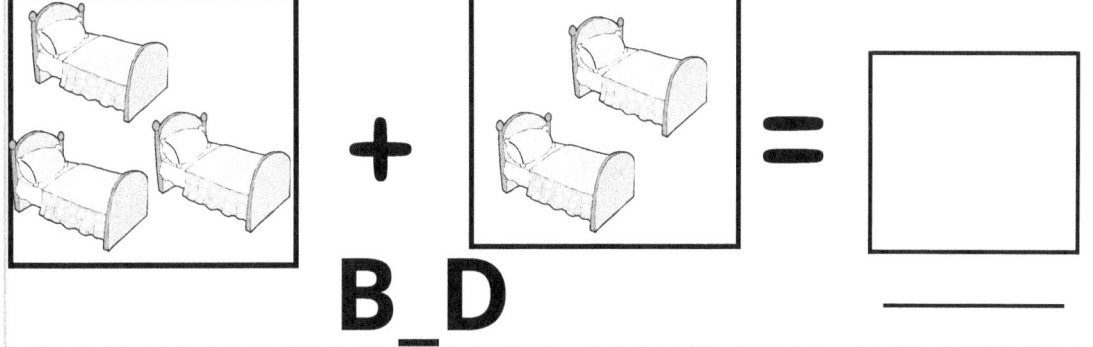

B_D

DE_K

SUBTRACTION

Subtract the number of clothing and write your answer in the empty box below. Fill in the blank spaces to complete the name of the clothing

| Name: |
| Grade & Section: |
| Age |
| Date: |

CLOTHING

TR_US_R

SH_ES

SK_R_

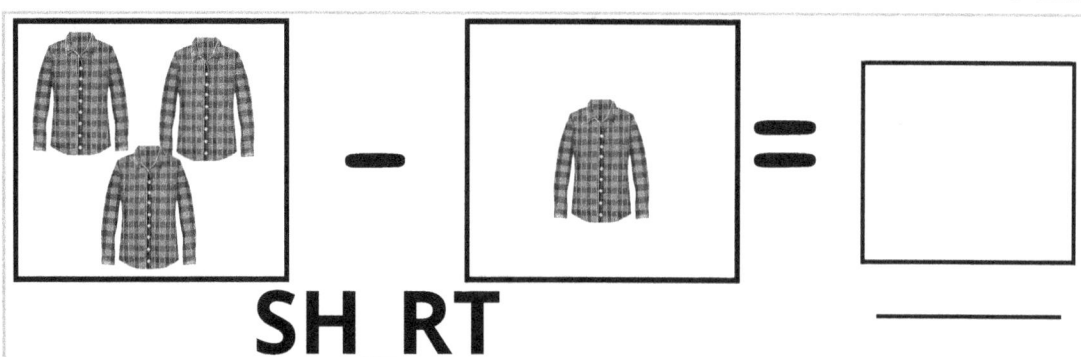

SH_RT

ADDITION

Write your answer in the box below!

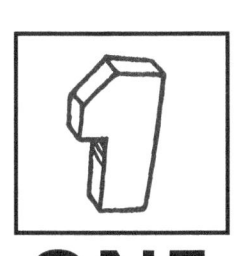 **+**  **=**

ONE   **THREE**   _____

 **+**  **=**

FIVE   **TWO**   _____

 **+** 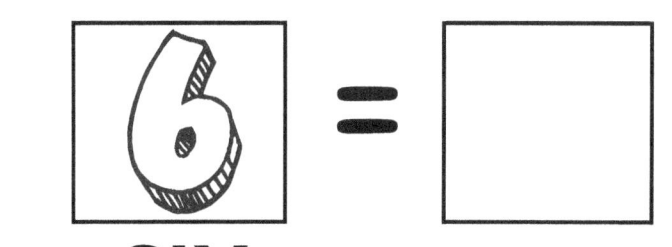 **=**

FOUR   **SIX**   _____

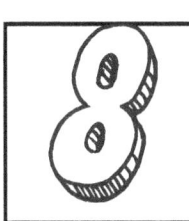 **+**  **=**

SEVEN   **EIGHT**   _____

SUBTRACTION

Write your answer in the box below!

FIVE — TWO =

FOUR — THREE =

SIX — THREE =

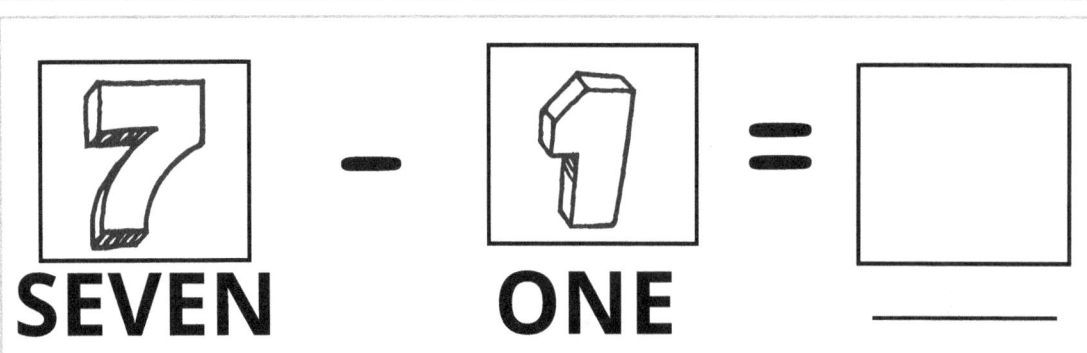

SEVEN — ONE =

Color the result

1+7= ○○○○○○○○○○

8+2= ○○○○○○○○○○

2+4= ○○○○○○○○○○

7+1= ○○○○○○○○○○

0+2= ○○○○○○○○○○

8+2= ○○○○○○○○○○

4+4= ○○○○○○○○○○

2+3= ○○○○○○○○○○

7+3= ○○○○○○○○○○

3+5= ○○○○○○○○○○

8+1= ○○○○○○○○○○

Color the result

3+7= ○○○○○○○○○○

4+2= ○○○○○○○○○○

4+4= ○○○○○○○○○○

3+1= ○○○○○○○○○○

5+2= ○○○○○○○○○○

7+2= ○○○○○○○○○○

2+2= ○○○○○○○○○○

2+3= ○○○○○○○○○○

1+3= ○○○○○○○○○○

3+4= ○○○○○○○○○○

3+1= ○○○○○○○○○○

Color the result

3+6= ○○○○○○○○○○

6+2= ○○○○○○○○○○

4+6= ○○○○○○○○○○

1+1= ○○○○○○○○○○

5+5= ○○○○○○○○○○

6+2= ○○○○○○○○○○

3+2= ○○○○○○○○○○

9+1= ○○○○○○○○○○

5+3= ○○○○○○○○○○

5+4= ○○○○○○○○○○

3+6= ○○○○○○○○○○

Color the result

7+1= ◯◯◯◯◯◯◯◯◯◯

5+2= ◯◯◯◯◯◯◯◯◯◯

7+0= ◯◯◯◯◯◯◯◯◯◯

2+5= ◯◯◯◯◯◯◯◯◯◯

3+7= ◯◯◯◯◯◯◯◯◯◯

4+6= ◯◯◯◯◯◯◯◯◯◯

7+1= ◯◯◯◯◯◯◯◯◯◯

4+5= ◯◯◯◯◯◯◯◯◯◯

9+1= ◯◯◯◯◯◯◯◯◯◯

8+0= ◯◯◯◯◯◯◯◯◯◯

4+6= ◯◯◯◯◯◯◯◯◯◯

Color the result

8+2= ○○○○○○○○○○

5+0= ○○○○○○○○○○

2+5= ○○○○○○○○○○

3+4= ○○○○○○○○○○

7+2= ○○○○○○○○○○

4+4= ○○○○○○○○○○

8+1= ○○○○○○○○○○

0+6= ○○○○○○○○○○

8+1= ○○○○○○○○○○

5+4= ○○○○○○○○○○

3+6= ○○○○○○○○○○

Color the result

8-2= ○○○○○○○○○○

5-0= ○○○○○○○○○○

5-2= ○○○○○○○○○○

3-1= ○○○○○○○○○○

7-3= ○○○○○○○○○○

4-2= ○○○○○○○○○○

8-2= ○○○○○○○○○○

9-1= ○○○○○○○○○○

8-5= ○○○○○○○○○○

5-4= ○○○○○○○○○○

6-3= ○○○○○○○○○○

Color the result

10-5= ○○○○○○○○○○

9-5= ○○○○○○○○○○

6-4= ○○○○○○○○○○

8-1= ○○○○○○○○○○

7-4= ○○○○○○○○○○

9-2= ○○○○○○○○○○

3-3= ○○○○○○○○○○

4-3= ○○○○○○○○○○

2-1= ○○○○○○○○○○

8-3= ○○○○○○○○○○

6-5= ○○○○○○○○○○

Color the result

10-7= ○○○○○○○○○○

8-3= ○○○○○○○○○○

7-5= ○○○○○○○○○○

3-1= ○○○○○○○○○○

5-4= ○○○○○○○○○○

6-2= ○○○○○○○○○○

9-4= ○○○○○○○○○○

3-2= ○○○○○○○○○○

8-6= ○○○○○○○○○○

9-3= ○○○○○○○○○○

7-2= ○○○○○○○○○○

Color the result

10-1 = ○○○○○○○○○○

5-4 = ○○○○○○○○○○

3-1 = ○○○○○○○○○○

7-7 = ○○○○○○○○○○

9-6 = ○○○○○○○○○○

5-1 = ○○○○○○○○○○

6-3 = ○○○○○○○○○○

7-3 = ○○○○○○○○○○

9-7 = ○○○○○○○○○○

4-2 = ○○○○○○○○○○

7-4 = ○○○○○○○○○○

Color the result

10-5= ○○○○○○○○○○

7-2= ○○○○○○○○○○

6-1= ○○○○○○○○○○

5-3= ○○○○○○○○○○

4-2= ○○○○○○○○○○

3-1= ○○○○○○○○○○

7-1= ○○○○○○○○○○

8-6= ○○○○○○○○○○

9-7= ○○○○○○○○○○

5-2= ○○○○○○○○○○

6-1= ○○○○○○○○○○

Let's do it.

8+2=	5+5=
5+0=	8+1=
2+5=	3+4=
3+4=	2+2=
7+2=	1+1=
4+4=	3+5=
8+1=	0+0=
0+6=	5+5=
8+1=	1+5=
5+4=	4+6=
3+6=	1+8=

Let's do it.

7-3=	4-4=
9-4=	9-7=
6-3=	8-5=
4-2=	5-4=
7-5=	6-4=
9-5=	9-7=
10-3=	0-0=
4-3=	7-1=
7-2=	3-1=
5-4=	7-4=
6-1=	9-3=

Let's do it.

7+2=	4-2=
5-2=	5+4=
6+3=	8-3=
9-5=	5+4=
4+4=	6-3=
8-2=	7+3=
10+0=	8-1=
7-2=	7+0=
2+5=	7-5=
9-8=	4+3=
1+5=	9-7=

Let's do it.

$4+1+5=$ $\qquad$ $4+4+1=$

$7+2+1=$ $\qquad$ $1+7+1=$

$1+5+2=$ $\qquad$ $5+1+3=$

$2+4+3=$ $\qquad$ $3+4+2=$

$0+5+3=$ $\qquad$ $4+2+1=$

$4+3+2=$ $\qquad$ $2+2+2=$

$2+4+4=$ $\qquad$ $1+2+7=$

$2+5+1=$ $\qquad$ $1+4+5=$

$1+1+1=$ $\qquad$ $7+0+3=$

$5+2+1=$ $\qquad$ $1+8+1=$

$3+5+1=$ $\qquad$ $3+3+4=$

Let's do it.

$9-7-1=$ $5-0-3=$

$5-3-2=$ $10-5-3=$

$8-7-0=$ $8-1-7=$

$5-2-1=$ $10-4-2=$

$7-4-3=$ $6-0-5=$

$8-4-2=$ $4-2-1=$

$9-5-3=$ $8-3-2=$

$4-3-0=$ $7-4-1=$

$6-3-1=$ $9-1-5=$

$7-1-3=$ $10-3-5=$

$8-1-4=$ $4-1-2=$

Let's do it.

4+1-3=	4-4+5=
7-2+3=	5+5-5=
6+4-5=	9-8+7=
10-7+5=	3+4-6=
2+6-5=	6-5+8=
8-5+6=	4+2-3=
2+4-3=	5-2+7=
8-4+3=	2+7-1=
8+1-4=	2-2+5=
9-7+5=	1+8-7=
7+2-4=	10-8+6=

MATH WORKSHEET
ADDING

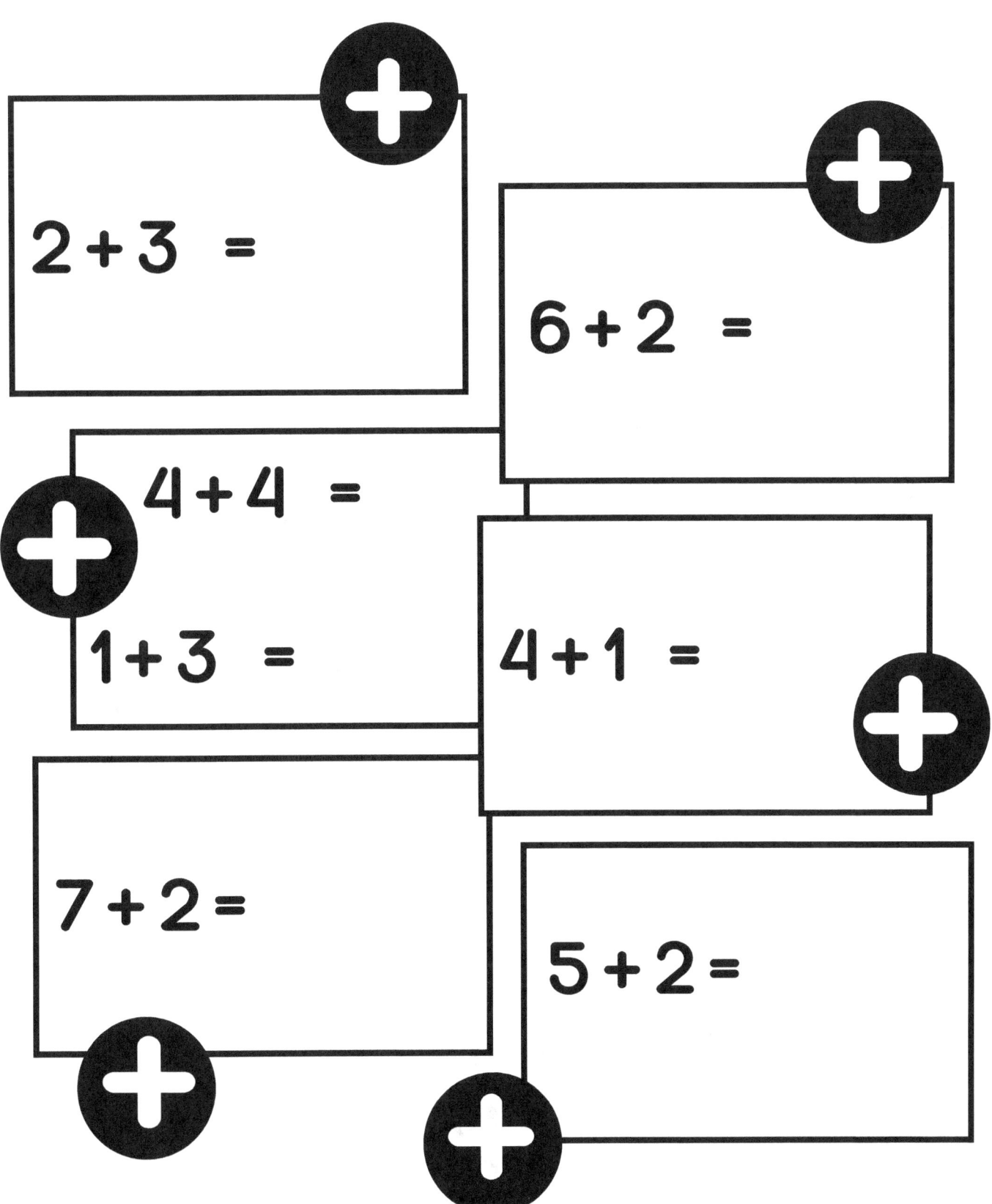

2 + 3 =

6 + 2 =

4 + 4 =

1 + 3 =

4 + 1 =

7 + 2 =

5 + 2 =

MATH WORKSHEET
ADDING

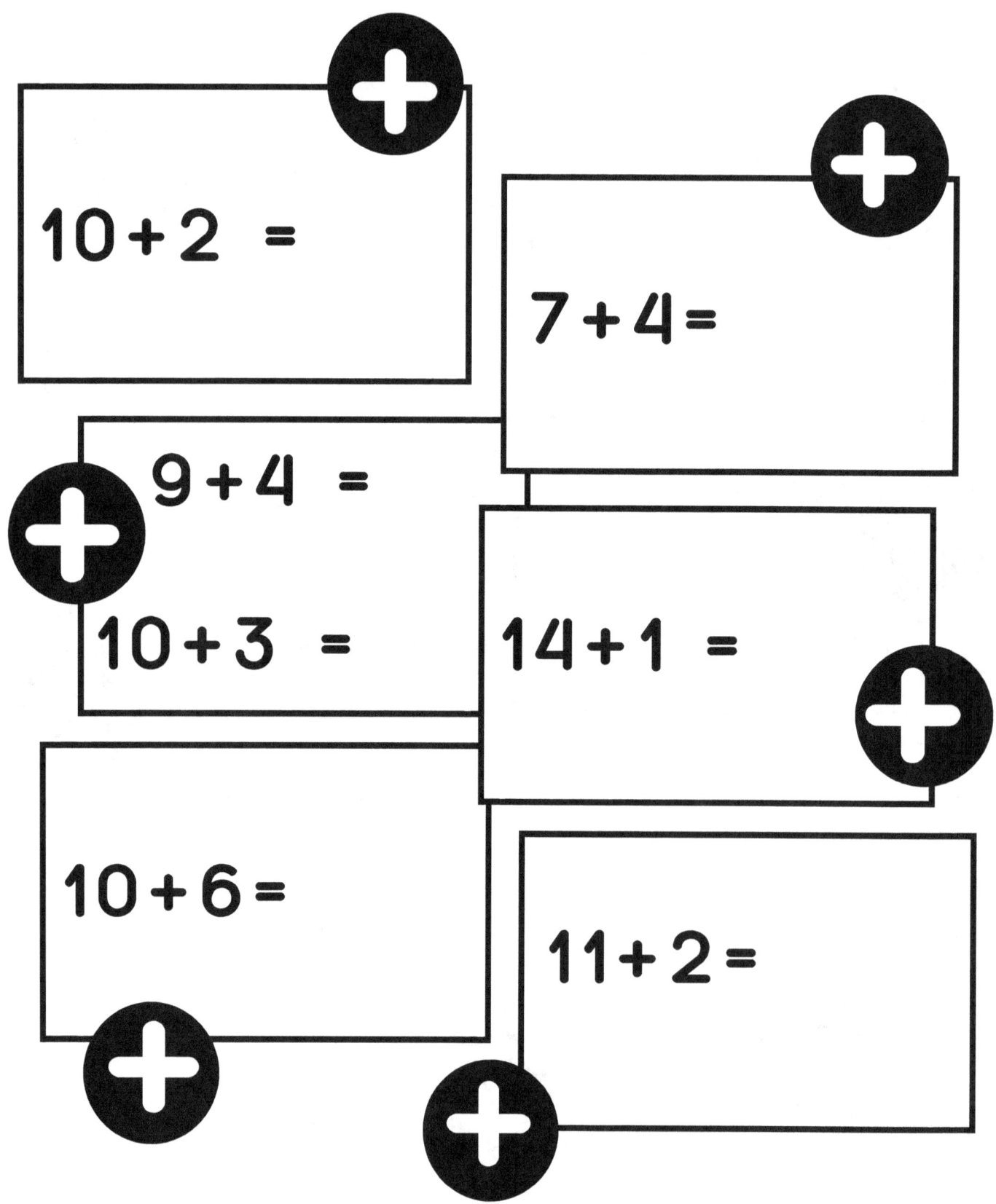

$10 + 2 =$

$7 + 4 =$

$9 + 4 =$

$10 + 3 =$

$14 + 1 =$

$10 + 6 =$

$11 + 2 =$

ADDING

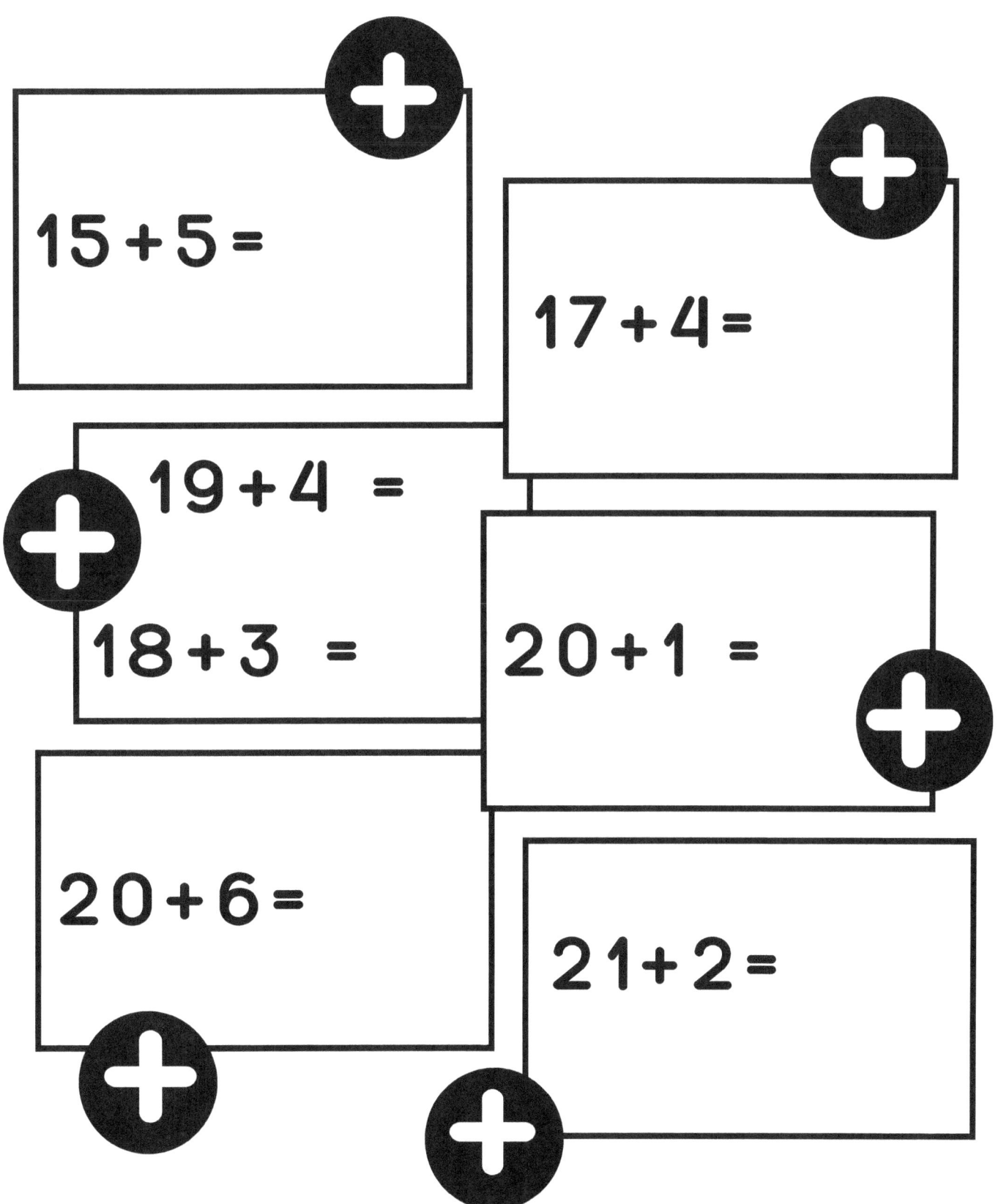

15 + 5 =

17 + 4 =

19 + 4 =

18 + 3 =

20 + 1 =

20 + 6 =

21 + 2 =

SUBTRACTING

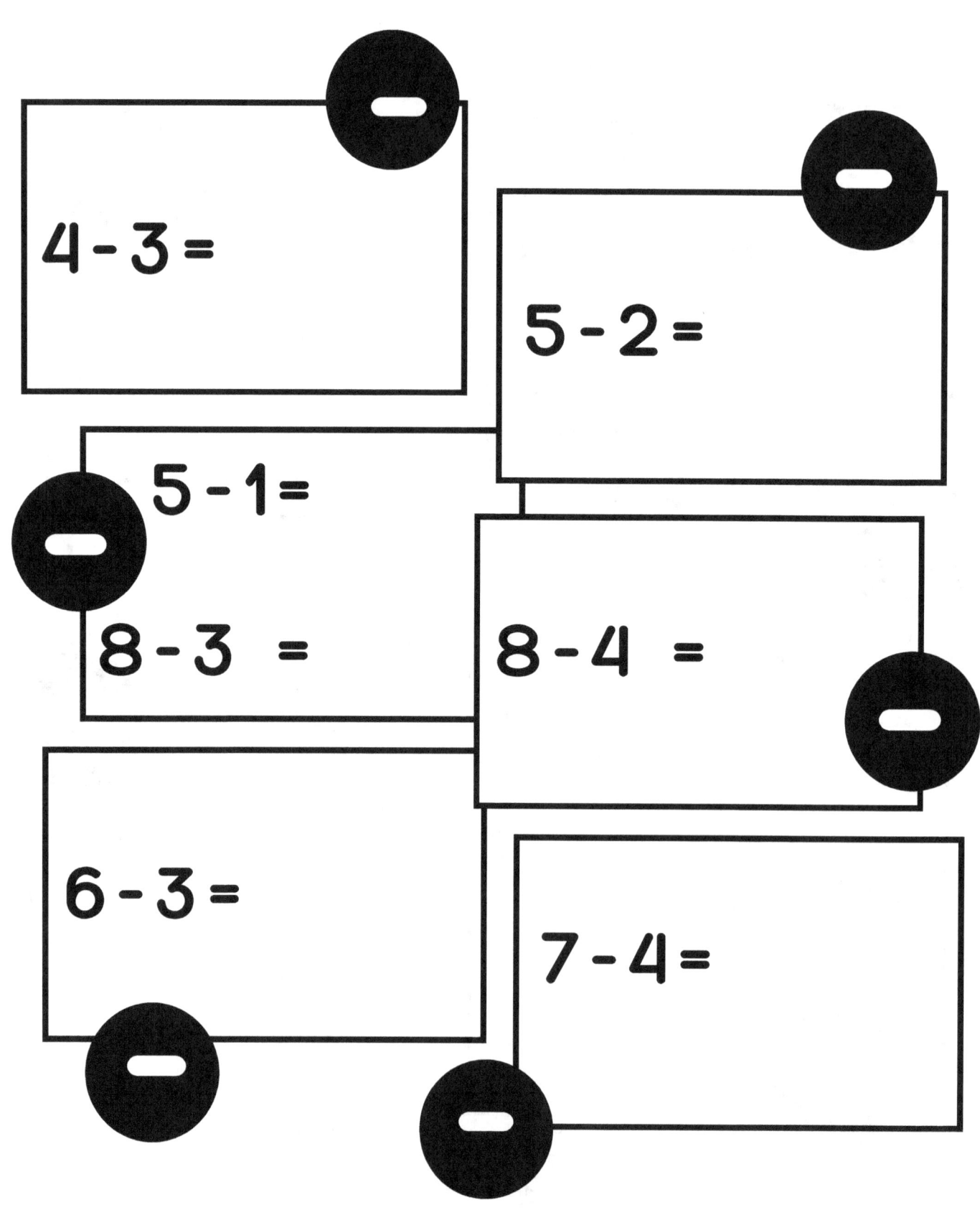

$4 - 3 =$

$5 - 2 =$

$5 - 1 =$

$8 - 3 =$

$8 - 4 =$

$6 - 3 =$

$7 - 4 =$

MATH WORKSHEET
SUBTRACTING

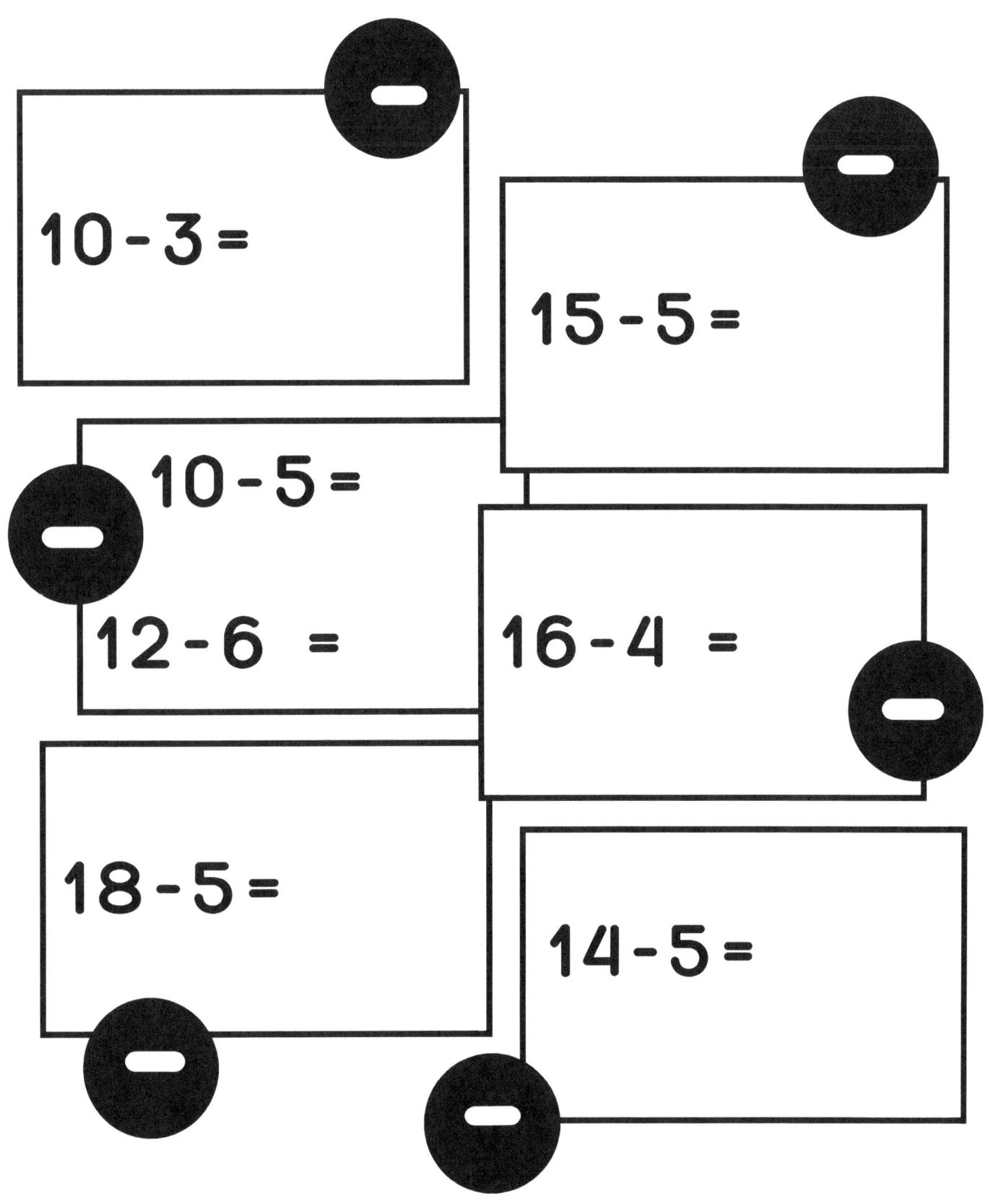

10 - 3 =

15 - 5 =

10 - 5 =

12 - 6 =

16 - 4 =

18 - 5 =

14 - 5 =

MATH WORKSHEET
SUBTRACTING

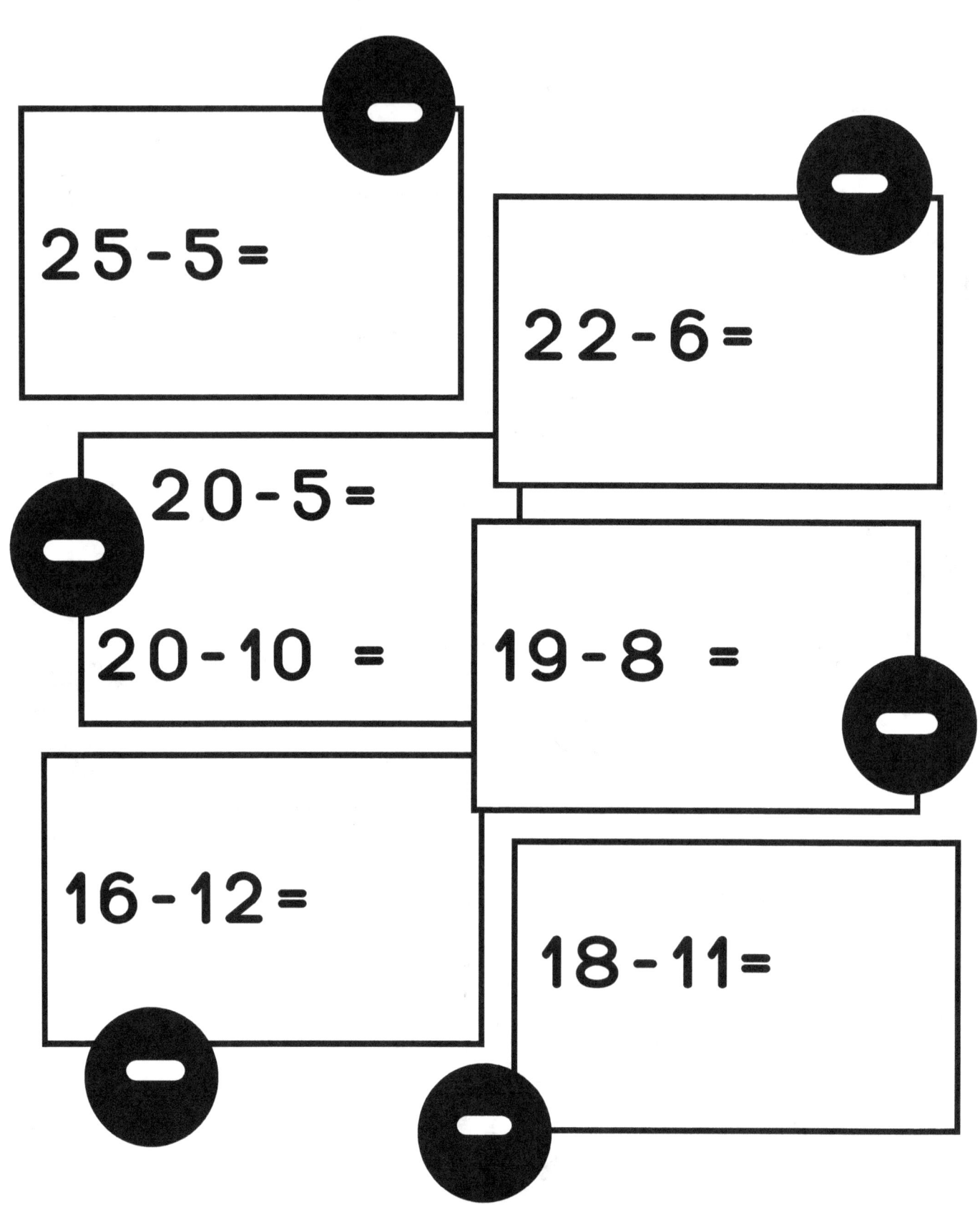

25 - 5 =

22 - 6 =

20 - 5 =

20 - 10 =

19 - 8 =

16 - 12 =

18 - 11 =

MATH WORKSHEET
COUNT THE OBJECTS AND WRITE THE NUMBER

MATH WORKSHEET
WRITE THE MISSING NUMBER

MATH WORKSHEET

FIND, COUNT, WRITE

MATH WORKSHEET
TRACE, COUNT, WRITE

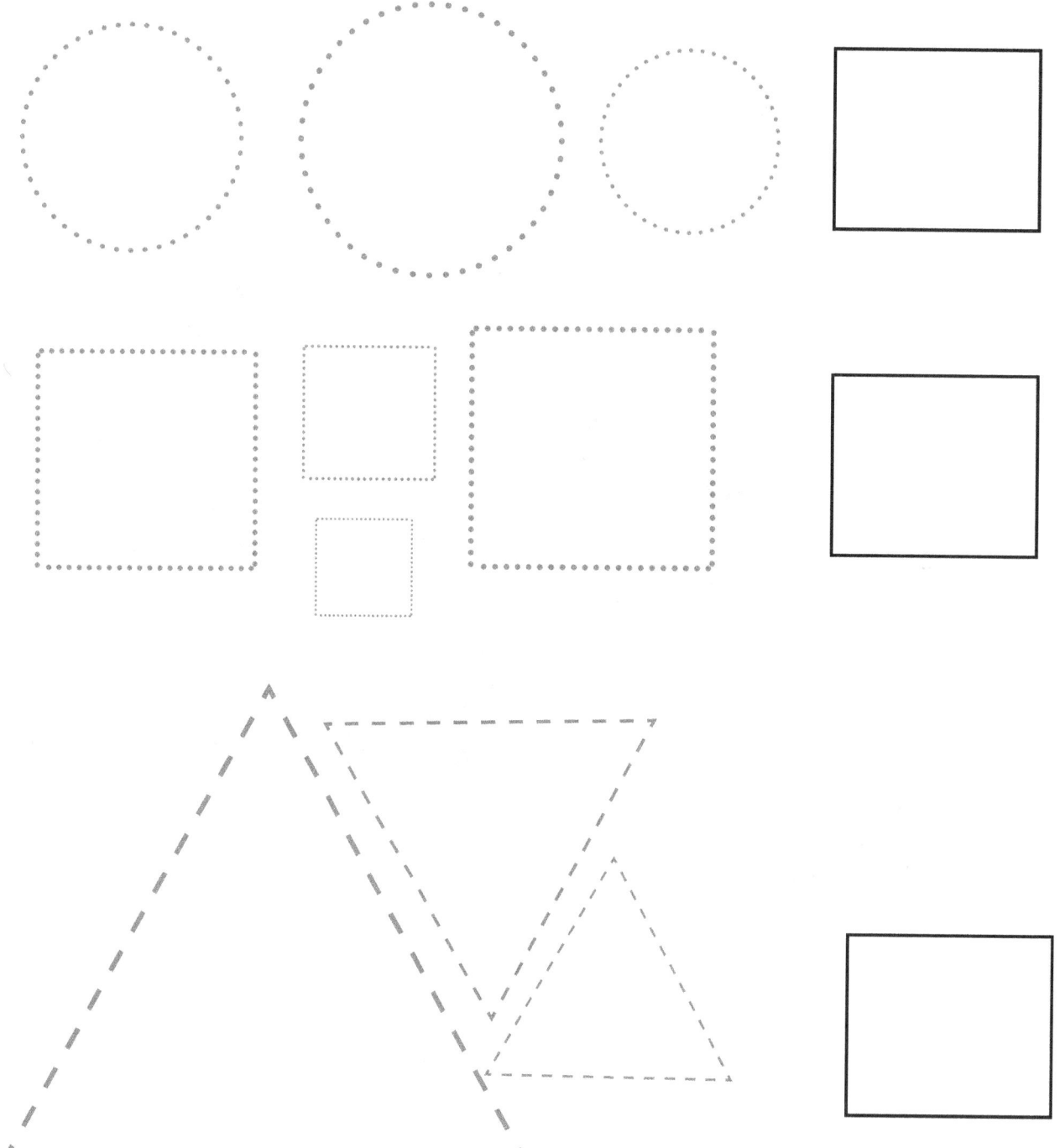

www.ingramcontent.com/pod-product-compliance
Lightning Source LLC
Chambersburg PA
CBHW080854220526
45467CB00008B/2511